Anonymous

## James G. Blaine on the Money Question

And Other Psychic Articles

Anonymous

**James G. Blaine on the Money Question**
*And Other Psychic Articles*

ISBN/EAN: 9783744759106

Printed in Europe, USA, Canada, Australia, Japan

Cover: Foto ©berggeist007 / pixelio.de

More available books at **www.hansebooks.com**

# JAMES G. BLAINE

## ON THE

# MONEY QUESTION

## AND OTHER PSYCHIC ARTICLES.

PUBLISHED BY

ÆTNA PUBLISHING COMPANY,

MINNEAPOLIS, MINN.

# INTRODUCTION.

Parting with friends is temporary death,
As all death is. We see no more their faces,
Nor hear their voices, save in memory;
But messages of love give us assurance
That we are not forgotten. Who shall say
That from the world of spirits comes no greeting,
No message of remembrance? It may be
The thoughts that visit us, we know not whence,
Sudden as inspiration, are the whispers
Of disembodied spirits, speaking to us
As friends, who wait outside a prison wall
Through the barred windows speak to those within.
                                        —Longfellow.

---

The message from James G. Blaine, herewith published, was received at nine sittings, by members of the North-western Society of Occult Philosophy, by means of a "Ouija" board, and was read at a recent meeting of that society.

The board differs from an ordinary "Ouija" board, in that it is larger, is a hard wood board, without paint or varnish, and, in addition to the letters of the alphabet and the numerals, has printed upon it some words most frequently used. The "traveler" also differs from the "traveler" accompanying an ordinary "Ouija" board. It has no legs, and is a piece of a cigar box, slightly warped, with a square hole cut in the center. It is laid upon the "Ouija" board with the concave side down, and, when in use, the hole in the center rests over the letter, numeral or word intended.

Early in January, 1896, by means of this board, communications upon philosophical subjects began to be re-received from a gentleman, with whom the persons receiving them had been acquainted, when he was in earth life. This gentleman was Frank G. Peters. Mr. Peters was a graduate of Yale College, a lawyer, and a gentleman of fine attainments. He was about thirty years of age when he passed over, which event occurred about three years ago.

One of Mr. Peters' communications follows Mr. Blaine's message.

These communications continued at sittings held regularly until February 9, 1896 when Mr. Peters withdrew in favor of Mr. Blaine. The last communication from Mr. Blaine included in the message was received March 9, 1896.

It will be noticed that there is a mistake of one day in the date of the "Act to codify the coinage laws," as given in one place in Mr. Blaine's message. The correct date is February 12, instead of 13, as shown by the record.

Mr. Blaine is probably not responsible for this mistake, and it was very likely caused by the fact that the numerals on the board are so close together that the hole in the "traveler" sometimes overlaps parts of two figures, making uncertain which is intended.

Mr. Blaine gives the correct date in another place in the message. He calls the act, "An act to codify the coinage laws." The title of the act is "An Act Revising and Amending the Laws Relative to the Mints, Assay Offices, and Coinage of the United States."

"The Act to Strengthen the Public Credit" and the contract between the Rothschild-Belmont-Morgan Syndicate referred to in the message will be found in the appendix.

The message is published more as a contribution to occult literature than as a financial argument, and is published at the urgent and repeated request of the sender. Those who believe that it is a genuine communication from the spirit of Mr. Blaine, will undoubtedly give it the same weight which they would accord to a like expression from Mr. Blaine if he were still living as a mortal.

As an argument it is not elaborate, and could not well be under the difficult and tedious method in which it was transmitted.

The results at the different sittings varied greatly, sometimes a comparatively long communication, and at other times a very short one being received. The inequality in the length of the communications may be attributed to difference in atmospheric, magnetic or electric conditions, or in the physical condition of the sitters, or possibly to a difference in conditions on the other side the veil. Towards the close of each sitting the "traveler" moved very slowly, showing that the power was being exhausted. On one or two occasions Mr. Blaine stated that he was compelled to stop by reason of the unfavorable conditions existing. All introductory and closing statements not part of the argument are omitted, and whatever is in parenthesis was spoken by some person present. One of the receivers of the message was, and still is, a believer in the gold standard, although he confesses that he is profoundly impressed by the message, and it is apparent that his faith is very much shaken.

The following resolution was adopted by the Northwestern Society of Occult Philosophy:

"Resolved, That we have heard with great pleasure the message from James G. Blaine, which has been received by means of a "Ouija" board, by some of our members; that in our opinion it furnishes very strong proof of the possibility and practice of spirit communion with mortals, and is both an important contribution to occult science and a profound discussion by a high authority of the most vital economic question before the American people."

# JAMES G. BLAINE ON THE MONEY QUESTION.

*A Message to His Country.*

## CHAPTER I.

Sunday, February 9, 1896.—(Mr. Peters, are you here?) Yes. (Will you talk for Mr. —————?) What does he want me to talk about? (Can you tell us anything more about God?) I have found no more than I had before. (Are spirits controlled by natural laws?) We are controlled by natural laws; everything is. I am forced to give way to another who has never before communicated with earth. He has a message of great importance for his country. (All right, give it.) By all that is good do not neglect this which I admonish you to do. The country is in peril; the monster which would tear it to fragments is abroad in the land. Every patriotic citizen must help or it will be too late. It is on the verge of destruction. (Who is writing?) JAMES G. BLAINE. (How are we to remedy matters?) Stop bond issues; the home loans especially. There is a dark and dangerous conspiracy to plunder the whole country. The East does not see it yet, and the West, with its better and quicker perception, is nearly as much in the dark. There is a plan to draw all the ready money from the channels of business in all parts of the country, and thus paralyze business. Under the guise, or better, disguise, of a popular loan they are drawing from hiding all the gold in the country, only to be shipped to Europe as soon as it gets to the United States treasury. That in itself is sufficient to condemn the whole transaction, because it depletes the very sinews of the country by such contraction of the circulation of the active and potential capital. It is further proposed to retire many millions of greenbacks by taking them in payment for these bonds and leaving them in the treasury, which will contract the currency that much more. The currency is now too low for the legitimate business of the country, and this course will be the last straw to break the camel's back. There have many excellent financial institutions perished during the last two or three years, which would have been all right today but for the unnatural condition of affairs. More money and not less is what the country must have, or every man, who owes a dollar, will be a pauper without hope, and the country will not be worth its bonds in the appreciated coin which is being paid.

I spoke of a dark conspiracy; I should have used the plural, for their name is legion. They are confined to no party, nor faction, nor country. They are being produced in all conceivable and inconceivable places. The great financial problem is but the result of many conspiracies.

The people of the United States have already paid the national debt several times. There have been schemes concocted, developed and executed, one after the other, by which under equivocal names and various pretences the debt has, or rather, the evidences of indebtedness have, been enhanced from time to time since the large amount of bonds were launched into existence in the early '60's.

As a result the debt is as large as ever and the payments have gone for naught. The national debt is larger today than it was at the close of the war. It will take more of the produce of the country to pay it now than it would have taken in 1865.* You are told that the reason for this state of affairs is because prices have fallen, but such is not the fact. The value of the dollar has been so astutely manipulated by these successive conspiracies, that the nominal decrease in the debt has been more than offset by the increase in the value of the money or units in which it is payable. As a matter of right the national debt has been more than twice paid; as a matter of fact, not a cent of it has been paid; and as a matter of prophecy, it will never be paid nor decreased as long as the bondholders control national legislation. The wealth of the fields, forests, mines and commerce of the country may pour in golden streams into this maelstrom forever, and it will still be as capacious, hungry and insatiable as ever.

There is a remedy, or at least a partial one, for the future. The past cannot be prevented nor its evils remedied. It is gone, and the future must profit by experience, or the vaunted freedom of American citizens will degenerate into the most abject slavery.

I spoke of the equivocal names under which the machinations of the conspirators cloaked their nefarious purposes. The most prominent act of Congress in the years shortly succeeding the close of the war, was called "An Act to Strengthen the Public Credit," passed March 18, 1869, which declared that all bonds should be paid in coin.* The

---

*See note 1 in Appendix.
*See note 2 in Appendix.

bonds had been purchased with paper which had fluctuated from 35 cents on the dollar to par, and the discount was the greatest when the debt was being contracted the fastest. The average price paid for the bonds in gold was but little more than fifty cents on the dollar. The act above referred to made the bonds worth two dollars for every dollar they had cost.

Another equivocal name for an act to disguise a sinister purpose was the act to codify the coinage laws, passed February 13, 1873, often referred to now as the "Crime of '73." By this act the value of the dollar was so manipulated that the bonds were again doubled in value, making them worth four times their cost price. The interest has nominally been reduced to an average of four per cent, but in fact you are paying four per cent on four times the debt you should, and paying it in a currency worth four times that in which the debt was contracted; so in fact you are paying sixteen per cent.

Stop for a moment and consider the position the country is in and the position it would have been in had the value paid on the national debt been justly applied as between man and man. The amount that has been paid in principal and interest is very near five billions of dollars, and in fact more than fourteen billions reckoned in the currency in which the debt was contracted. What is the situation today? You have paid as much as the total valuation of all property in the United States in 1880, both real and personal, and you still owe as much as you did at the close of the war. Had this money which has been paid been properly and justly applied, the nation would now be free from debt and the balance that has been paid above what was necessary to discharge, the nation's obligations would have purchased every railroad, telegraph and telephone line in the United States, and would have dug every canal. ever proposed in the country and the Nicaraguan to boot.

Numerous other acts have been passed at various times, such as refunding laws by which the bonds were given a new lease of life for a long period at an apparently lower rate of interest; and right here I wish to let you know of a plot that is now being concocted in the same direction, but by means never before resorted to. I told you the first time I was with you to stop the popular loans. The capitalists of Europe are cornering the gold. After getting legislation making nearly everything payable in gold and by collusion with executive officers making every obligation payable in gold by construction, they now have the matter where they can handle it, and can now proceed to draw the gold to Europe and store it until the next bonds are due, and as there will not be any gold in the country with which to pay them, they will force the government to again extend the period of payment for a long term of years. The popular loans are simply to pick up all the loose change in the way of gold that is scattered through the country and get it into the United States treasury, whence the road to Europe is a steep down grade.

## CHAPTER II.

The remedy for the present condition of affairs is what you are looking for, but there is no reason for being too precipitate. You would better first consider the consequences in case you apply no remedy, so you may more fully appreciate the gravity of the disease you are to remedy.

I will tell you of a small-sized conspiracy which is being consummated at the present time. It is a wheel within a wheel, and though it is large enough, I call it small, because it pales into insignificance, when compared with the great or principal plot.

It is in connection with the recent bond sale by means of sealed bids. There were nearly ten millions of bonds awarded at a fraction over 115, which were defaulted by the bidders failing to make the deposit of their money as required by the regulations of the secretary, and those bonds were given to the Morgan syndicate at a fraction over 110. The difference in the amount received by the government and what it should have received is about $500,000. This was planned in advance, and the parties whose bids for the bonds were accepted and who then defaulted, will get the same bonds from the Morgan syndicate and will divide the difference.

There is no necessity for drawing this tale to a great length. The country has poured its wealth into the pockets of greedy bondholders until its resources have become decimated, the people have become discouraged, and the ambition, which formerly characterized the American people, has nearly disappeared. The country, richest in natural resources and richest in its citizenship, has fallen before an invisible foe which worked in the dark, boring slowly but surely toward the heart of this puissant nation. A few marks of the pen have accomplished that which all the armies and navies of the world could not have done— another illustration of the pen being mightier than the sword.

If the country had some business men at its head instead of politicians, there would be some hope for its recovery. There is much that can never be remedied; but if the affairs of the nation could for a time be run as a private citizen would run his business, there would soon be an end for its financial ills. No country has such recuperative powers and possibilities as the United States of America.

You must do some figuring for yourselves. Do not believe when you are told that only financiers with long experience can understand the national finances and how to manage them. If you do believe that, what do you think of the chances of the country at the present time? Where are the trained financiers in this administration? The people must figure for themselves. Others have been figuring for you too long and have figured the country where Tommy did the frog.

The incident to which I refer occurred in a small town not far from Bangor, Maine. A small boy in a rural district was given an example to figure out by the teacher. She said: "Now, Tommy, if a frog is in a well and he jumps up one foot every day and falls back two feet every night, how long will it take him to get out of the well?" Tommy figured for about two hours and covered his slate and that of his seatmate with figures. "How are you getting along, Tommy?" asked the teacher. "Oh," replied Tommy, "I've got him half way to hell already."

Those who have been doing your figuring for you have reached about the same result, and the more they figure, the nearer they bring the country to ruin. Therefore, I repeat, you must figure for yourselves. Go at it as a merchant or any other business man would. The bonds of which I have previously spoken are payable in coin, and by coin is and was meant the coin that was in use or was authorized by the laws in force when the bonds were made. The metal of one of the coins known to the laws at the time the bonds were made can now be purchased in the markets of the world for about half the cost of the other metal, comparatively. If a business man owed a debt and had the same option in regard to payment that the American government has, which metal would he pay it in?

Does it require a trained financier to answer that question? Does it even require a business man? Cannot a farmer or a mechanic, or a child give the correct answer as well?

The officials at Washington are but the agents of the people, and as such are bound to do the best they can for their principal, the people of the United States.

Suppose you owned a farm and hired an agent to look after your interests in connection therewith, and suppose you owe a debt of a thousand dollars which your agent is directed to liquidate from the crops. Suppose the debt is payable either in gold at 23.22 grains to the dollar, or in silver at $371\frac{1}{4}$ grains to the dollar, and supose one bushel of wheat would buy sufficient silver to pay a dollar, and it took two bushels of wheat to buy gold enough to pay a dollar, and suppose your agent paid in gold, how long would you keep him after you found it out? That is exactly what your agents at Washington are doing; they are taking nearly twice as much of your property as is necessary, with which to pay your debt.

The illustration of the farmer and his wheat applies as well to the merchant, manufacturer or any business or profession.

At the present time, fifteen yards of calico, at the mills in Massachusetts, will buy 371¼ grains of pure silver, and twenty-five yards will scarcely buy 23.22 grains of gold. Every time a payment is made in gold, property equal to the amount of the payment, and which belongs to the people of the United States, is confiscated and given to the bondholders. If payment were made in silver, the bondholders would get back more than double the value they paid. Is it just to confiscate the property of the country as above shown? Would it be unjust to pay the bonds in a metal worth twice as much as the currency in which they were purchased?

I think any one can see what I mean, but for fear they may not, will say that the business of the country, should be conducted on business principles. When an obligation is presented for payment, pay it as cheaply as you can. If the silver in a silver dollar is cheaper than the gold in a gold dollar, and the debt is payable in either, pay in silver as long as a fraction of a mill can be saved. Rothschild would do so to you; do you even so unto him.

## CHAPTER III.

I will say a few words giving my present convictions on the question of the free coinage of silver. I will begin by stating that I am unqualifiedly in favor of the free, unlimited, immediate and independent coinage of silver; and I will tell you why. It is the constitutional money of the country. It has been the money of the world for nearly three thousand years. It was the money of redemption when the present debt of the country was incurred, which made every inhabitant of the country a debtor. The demonetization of silver took away one-half of the redemption money and doubled the value of what remained; it practically confiscated one-half of the property of the United States.

Your property is being swept away to purchase appreciated gold dollars to pay a debt contracted when gold, silver and paper were legal tenders, and silver enjoyed the same privileges at the mint that gold did. One by one the legal tenders have been eliminated until there is today practically but one, and that is in the hands of a few who have the power to contract it at will and who are actually cornering it at the present time. If you have any doubt about their ability to control the gold of the world, read the contract entered into in February, 1895, between the United States of America, party of the first part, and the Rothschild-Belmont-Morgan syndicate, parties of the second part. That was made for the protection of the United States treasury, and is the only protective instrument on record that President Cleveland ever favored.*

I shudder to contemplate the result to the republic if the great wrong done by the act of February 12, 1873, is not remedied, and that at once. Delay is more than gerous, it is perilous. Unless everything possible is done, and at once, there will be such a revolution in the United States as the world never saw. Already the mutterings can be heard, and I will predict that unless remedial legislation is had within two years, the revolution will be upon you in all its fury.

The American people are brave and long-suffering, but

---

*See note 3 in Appendix.

when they see the accumulations of a lifetime ruthlessly stolen from them and given to heartless bondholders, they will not bear it with equanimity. The prospect of laboring for a lifetime and having nothing to leave his children but poverty and debt is not a pleasant prospect for the average citizen of a country like the United States. The first thing that should be done is to restore silver to the place it occupied for over eighty years in the coinage and currency of the United States.

War is justifiable when peace would be more disastrous; and peace on the present terms would be more than disastrous, it would be suicidal.

I have no more to say. I wish you would say to the people among whom I lived and to the country which I loved, that I still live, and still love the United States of America. Tell them for me that if they love their country, if they prize their liberty, if they value their property, if they cherish their families and would protect their homes, they must at once undo the terrible injustice perpetrated on the country and its people by the demonetization of silver.

I have no more to say, at least at present; I simply wish this message delivered.

# COMMUNICATION FROM FRANK G. PETERS.

Good Evening: I am engaged in studying life; the vital principle; and I find myself engulfed in a labyrinth, more intricate, devious and insidious than the one in which Theseus sought the Minotaur. In the whole thesaurus of human knowledge there is not a word or syllable which throws the slightest light on this greatest of subjects.•

Is it because the subject is so vast that the intellect cannot compass or conceive it? Is it like space and duration, and should it be added to that category? Or is it so simple that we cannot comprehend it? I cannot give any very satisfactory results as yet, but I have found that the vital principle is the same in animal and vegetable life. If you cut your arm it will heal and leave a scar. If you cut a gash in a tree, exactly the same result will follow. A chemist can tell exactly the amount, by weight and measure, of every element in a tree or a dog, but he cannot make a live tree or a live dog. The vital principle may remain dormant for centuries. For instance, when Herculaneum and Pompeii were excavated, after having lain under the lava of Mount Vesuvius for nearly eighteen centuries, wheat was found, which upon being planted grew as well as new wheat.

Is life but a synonym for power and are the phases of life incidents or expressions of omnipotence as knowledge is of omniscience? Think of the power in one little seed. Take a mustard seed for example; it is one of the smallest of seeds: it grows and makes a tree and produces thousands of seeds capable of doing as much as the original. Think of the weight a mustard tree will support; then multiply by the number of seeds, which one seed will produce, and you find a mighty power. I am not much given to quoting scripture, and I may make a mistake, but somewhere it says, if you have faith even as a grain of mustard seed, you could say to yon mountain, remove hence into the sea, and it would get there. If you substitute the word "power" for "faith," it would be sensible and according to the fact.

Take a tobacco seed; it grows and produces a large plant and thousands of seeds like itself. The original seed contained a small particle of nicotine; the leaves contain nicotine, and each seed contains as much as did the original. There is more nicotine than there was before. We were taught that all the accretions to the plant life came from the atmosphere and from the earth, but what chemist ever said that nicotine was in either?

I will write more another time. Good night.

## LETTERS FROM THREE FAMOUS PSYCHICAL RESEARCHERS.

Eastnor Castle, England, August 7, 1893.

Professor Elliott Coues, President World's Psychical Congress—Dear Sir: I am glad to have the opportunity of sending my greetings to the Congress. I trust its deliberations may be beneficial to humanity in the present stage of the great and growing controversy between the spiritual and the material philosophies. In my opinion a clear distinction should be drawn between the interrogative temper of mind in which a scientist approaches the study of nature and what is called supernature, and what may be designated as the affirmative temper of the credulous. The danger lies in a too ready acceptance of what appear to be genuine manifestations, but which are in a great majority of cases illusive and delusive if not demoralizing. For this reason a scientific Congress will be of inestimable value, and I wish it success. Believe me, yours very truly,

—Isabel Somerset.

Professor Elliott Coues, Etc.—Kind Friend: For many years I have belonged to the American, and I have recently joined the British, Society for Psychical Research. For I have never been one of those who hold that there are subjects which we are forbidden to investigate; indeed, such a position involves, to my mind, nothing less than downright superstition. If man's reason and nature's phenomena are to be kept apart at any point, then why not at many points?

Whatever exists in the universe is a legitimate subject of thoughtful and reverent study by "man's illimitable mind." For this reason I have always been sympathetic toward the investigation, from a scientific point of view, of all psychical phenomena. I do not approve of elevating these investigations into a form of cult, any more than, for instance, those of astronomy. As a devout disciple of the Founder of the Christian religion, I would not take the positions herein stated did I deem them inconsistent with the gospel declaration that we are to "prove all things" and "hold fast to that which is good."

Believe me, with respectful salutations to the Congress,
Yours with best wishes,
—Frances E. Willard.

---

7 Kensington Park Garden, London, July 27, 1893.

My Dear Professor Coues: .... If you hear any rumors that I have backed out of the subject because I have found out that I was taken in, or in some other way found reason to disbelieve my former statements, you have my full authority—nay, my earnest request—to meet them with my full denial. As far as the main facts and statements I have recorded in the different papers I have published on the subject of the phenomena of Spiritualism, I hold the same belief about them now that I did at the time I wrote. I could not detect at the time any loop hole for deception in my test experiments; and now, with the experience of nearly twenty years added on to what I then knew, I still do not see how it was possible for me to have been deceived. Read my recently published "Notes of Seances with D. D. Home," and the introduction to these "Notes," and you will see what my present attitude of mind is.

With kind regards, believe me, very sincerely yours,
—William Crooks.

# TRANSCENDENTAL PHYSICS;

## Or Scientific Sources of Spiritual Truth—A Classical Essay on Some Fundamental Principles, by Solon Lauer.

Over the clouds of scientific agnosticism is rising the bright sun of a new faith, which shall illumine the world as no faith has ever before illumined it. Science, which is only the orderly observation of phenomena and the orderly classification of obtained knowledge, is coming to the aid of faith, and proving that the world-old beliefs of man concerning the soul and its divine nature and powers are not mere superstition, but veriest fact. Science, which began by doubting and even denying spiritual truth, has come to be its chief witness and its warmest friend.

In the first place, science has shown us that every form of matter can be reduced to an invisible vapor; that the solid world and all therein came forth out of invisible substance, and will sometime return to it again; thus verifying the old saying of Paul, that the things that are seen are temporal, but the things that are unseen are eternal. A rock, a piece of steel, seem substantial, and much more real than a principle of mathematics; but chemistry can reduce these to an invisible vapor, which at once eludes all our senses; while the mathematical principle remains forever in our mind.

When we have once perceived that the physical world is transient and fleeting, but that among all the changes of matter there is something which does not change, namely, mind, we are ready to consider the possibility of many instances of the supremacy of mind over matter which have come down to us in the traditions and writings of the world's great religions; for these are not peculiar to Christianity, but are found in every religion. We might go through the gospel records, and find that most of the miracles recorded therein of Jesus and his disciples are rendered creditable by facts of present experience. The discoveries in the realm of psychical research have thrown a great light upon the Scripture records.

The records of the London Society for Psychical Research are full of instances of occult power, observed by men and women of unimpeachable character and of the highest scientific reputation. Thought-transference is to-day a fact as well established by scientific evidences as any fact of psychology; and he who doubts it betrays either a gross prejudice or an unpardonable ignorance. It has been demonstrated beyond doubt that thought is a mode of force; that it goes out from the mind in waves of vibration, just as the undulations of heat and light go out from a lamp or stove, that it travels not alone over the channels of the nerves, but that like the force of a magnet it goes out in invisible streams to affect other minds within the sphere of its influence; and that so far from being limited to the physical agents of the body, mind has a power to transcend space, to act where the body is not, and to achieve results after laws of its own, which, though yet but little understood, are nevertheless as capable of study and demonstration as are the laws of electricity and magnetism, which are as yet but in the infancy of their development. When we investigate these laws of the mind, we find that they make clear many of the miracles reported in the gospel; and so far from making them less wonderful, render them more so, by making them a part of that great mysterious nature whose laws and phenomena, however much investigated, must yet forever remain a source of marvel and mystery to the human mind.

We find that among the phenomena of the mind which science has investigated and declared real and true there are many which so exactly parallel the reported miracles of the Bible and other religious records, as to leave no doubt in rational minds that the same laws prevail in both cases. For instance, thought-transference is often manifested by Jesus and the disciples, in a manner similar to that observed in these days. When Jesus told the women of Samaria all that ever she did, he exercised a gift which is not exclusive with him, but common to many in these times. Through the law of mental sympathy, he became aware of her past history, and read it as from an open book. The law which rendered this possible is a law of nature, a law of mind, and, supernatural as it may appear,

is nevertheless as natural as the law of vibration in light
or the law of induction in electricity.

The miracles of healing reported in the gospel records
have been a source of much controversy among commen-
tators, and it is common to hear their truth denied by liberal scholars. This denial takes from the gospel its very
life and spirit, for when the soul is robbed of its belief in
spiritual powers, religion soon ceases to be a thing of joy
and necessity, and becomes a mere ornament to society, a
sham, a farce, whose creaking formality is painful to every
earnest soul. There is another form of skepticism, however, which is equally baneful, and that is the form found
among many, probably most followers of orthodoxy in
religion. These accept the miracles of healing reported
in the gospels, but deny that any such are possible today;
thus cutting off the present believer from those benefits
enjoyed by the believers of early time, and promised to
all who should believe. We need to use caution in the investigation of so-called mental cures, but there should no
longer be any doubt among honest minds that such cures
are achieved. Under the various names of Christian Science, mental science, spiritual science, psycho therapeutics,
suggestive therapeutics, etc., cures have most certainly
been wrought which make easily creditable most of the
so-called miracles of cure wrought by Jesus and the disciples. We do not yet understand these laws of mind which
work such potent results in the realm of the body, bringing health out of disease and sometimes seeming to snatch
the departing spirit back from the very door of the tomb,
but we understand them as well perhaps as we do the laws
of electricity, which, though a deep and deepening mystery to all students, have yet been applied to human needs
so as to vastly lighten human labor and increase human
comforts. It is not long since the phenomena of electricity
were practically unknown. The attraction of a piece of
rubbed amber for straws and other light objects, the attraction of the loadstone for iron and steel, were not so
many years ago the sum of human knowledge upon this
vast subject. Even today we do not know what electricity
is, we do not know its nature, we know only the conditions
for certain of its manifestations. Our knowledge of the
mysterious laws of mind will certainly increase as our
knowledge of electricity has done; and the far-seeing soul

can perceive the time when, through the well-understood laws of mind, disease shall be cured without the use of drugs, vicious habits shall be reformed without the employment of brute force, and human society shall reflect something of that glory which has been sung in the lines of poets and chanted by the lips of prophets since time began.

The value of these recent researches of science to the Church and to religion is apparent to every one who will observe the signs of the times. The decay of belief in the spiritual phenomena of the Bible is too apparent to need more than statement. So-called liberal thinkers on every side deny them. Orthodox believers accept them as supernatural, limited to the days of the apostles, having no value to us today. This unbelief, as I said before, is as baneful as the other; for it makes human life today poor and wretched, with no possibilities of contact with those higher realms of life which made Christianity the redeeming force it was in the corrupt times of the Roman Empire. It is not until we realize that God is not the God of the dead, but of the living, that our theism has any vital force in our own life; it is not until we realize that whatever miracle was possible to Jesus and his disciples eighteen hundred years ago is possible to mankind today, that our belief in Christianity is of any practical value to us today. It concerns us not what has been done; but what may be done today; not what sick man of Palestine was cured, but what sick man of our own acquaintances may be redeemed to his natural right of perfect health. The wheat that grew on Judean hills will not feed the hungry of today. There is wheat growing in our fields, corn in our valleys. The bounty of God is never exhausted. It serves the present as the past. Man's relation to God is the same today as in the days of Christ. What laws prevailed in his life were laws of nature, and prevail today. Our insight of nature's laws must be so deep that we can include all truth and all real miracle in this realm of law and order. Nature is infinite, the manifestations of an Infinite Mind. Her laws are revelations; her phenomena miracles evermore; and as we press farther into her charmed realm, our wonder ever grows and deepens. It is not faith alone that can save us; but knowledge as well; for true faith is not a belief, but a perception. The Church today is dying of conventionality and unbelief. Her sacred aisles re-echo to the frivolous voices of men and women who have no sense of the sanctity of the place. Her pulpit resounds with platitudes of literary culture, with unctuous flatteries of the rich and delusive doctrines of vain Utopias for the poor. The stern voice of the prophet is seldom heard. Ideal living, spiritual culture, the opening of the soul's eyes to catch glimpses of that pure life which is man's true estate, these are almost unknown in the modern Church; but instead of spiritual rhapsodies and ecstatic visions and glad chaunts of divine love and life, we hear empty laughter, frivolous conversation, the clatter of dishes at Church suppers, the clink of moneys at the tables of the Church money-changers, and such other sounds as befit the public market better than the holy place of God. The sick are left to the tender mercy of physicians, whose drugs are believed in above the power of the spirit of God, which created drugs and physicians alike. Physic is more potent than prayer, though history is full of the triumphs of the latter. The lancet and the leech have retreated before more humane methods of treatment, but medicine is still a secular science, not baptized with the spirit of the living God. It is for the Church today to recover those spiritual verities which have been the life of religion in all ages, and make her sanctuary a holy place indeed, a sanctuary of the Divine Presence. Too long we have sought for religion where it abides not. It hath its temple in the devout mind, in the soul at one with God; and its revelations are words of souls smitten with the vision and ecstasy of the Divine Presence in human life. Let us do somewhat to recover this pristine glory of the Church; to restore religion to that divine estate which is her due. Our modern faith must be based upon modern revelation; our miracles must be miracles of living men and women, not records of the dead past only; our prayers must be the utterance of our own faith and perception of spiritual laws; the claiming of our divine heritage, which we have forgotten or sold for some wretched mess of pottage. Then shall the Church be once more the abiding place of God; His presence shall shine round about us, filling our lives with light and joy; sin and disease shall fade away before that Presence, as fog before the rising sun; and we shall live now in that kingdom which is promised to the saints, a kingdom of peace, joy, and good-will among men.—*Light of Truth.*

# HOW LITTLE WE ALL KNOW.

*Humanity's Newest Glimpse of Infinity — Beauties All
About Us Which We Cannot See — Music to
Which Our Ears Are Not Attuned.*

We read and hear a great deal nowadays about the
Roentgen ray. As I understand it, when a ray of sunlight
is separated by a prism into the different primary colors,
the spectrum has always shown also, both below and above
the colors, other lines which have not been designated and
which produce no effect upon the human eye. These unknown rays are found to possess the quality of passing
through what has been estimate as solid substance, and photographing more solid substances like the
skeleton of the human body, the adipose tissue showing
a mere outline. It is not photography exactly, the object
standing between the rays and a sensitized plate and the
picture being taken much as a blue print is in an engineer's
office. Is there anything substantial? Is everything merely
a form of motion? There is sound, heat, light, electricity,
each form of motion of greater or less intensity. We say
that we speak. A vibration of air through the vocal chords
sets in motion the sound waves which, striking upon the
drum of the ear and being communicated by a curious
mechanism to the brain, gives a sensation which we call
sound. It is nothing but a vibration of the air, and to
those whose ears are out of order there is no such
thing as sound. We say we see an object. From that object comes a light vibration which produces a picture on
the retina of the eye and there it is upside down, and in
some way, being communicated to the brain, we say that
we see a tree, a house, or a friend. Do we see anything?

If the eye is out of order these things do not exist so
far as the blind one is concerned, except the person becomes conscious of them through some other sense. Take
all the senses away and there is nothing but a clod. Blood
may circulate in it and it may breathe air, but it is only
a thing. Now, does not this take the conceit out of us to
some extent? We are so proud of what we know, which

has all been obtained through the use of the senses. And
we think we know everything and we say we will believe
nothing except that which we can see and hear and touch,
and yet how much is there of that which exists which we
may become cognizant of by the use of our senses? Each
new discovery, like that of this Roentgen ray, shows how
much there is around us of invisible forces which we have
not yet grasped. We have, to be sure, in the course of
centuries, gained considerable. We have found out something of the effects of electricity, although we do not know
what it is. We have done marvelous things with the sunlight, although it is now shown that we have not known
much about it, and so we are beginning to discover to
what a small limit our human powers are hedged. We
sneer at the miracles and say that they are a violation of
law and could not possibly have occurred, but what do we
know of natural laws? Almost every day some new thing
comes up which shows conclusively that a law existed of
which we never knew before. How do we know that
Infinite intelligence, which knows all things, could not
set in operation forces of which we have not the slightest
idea? "There are more things in heaven and earth,
Horatio, than are dreamed of in your philosophy." There
are objects all around us which we cannot see, the air is
filled with fragrance which we cannot smell, the universe
is crowded with sounds which we never hear, neither do
we know that anything looks the same to any two of us.
I say, by a certain sensation from a lady's dress that it is
blue; you also say that it is blue, but it is only by common
consent. The sensation which I call blue may not be at
all the sensation which you call blue. If we could exactly
get at the real truth we might find that we were far apart,
but having agreed to call in each of our minds a certain
sensation received through the eyes blue, by comparison
that always remains blue to us. Some people's eyes are
defective and do not separate these light rays and we call
them color blind. They cannot distinguish red from blue.
It looks all the same to them. And again, there are some
whose eyes have become so educated, like those of the
Florentine workers in Mosaic, or some East India workers
in tapestry and rugs, that they can distinguish as many
as 70 different shades of red and match them and differ-
entiate them. We know that a dog's sense of smell is such

that he can distinguish a scent where the human being
would not be able to, and can follow the trail of a person
uneringly over miles of ground.  A cat can see in the
night where the human eye can distinguish nothing.  You
see our powers are limited.  We do not know much, after
all.

Now, take this Roentgen ray, for instance, which is beyond the power of human eyesight.  If we had that capacity
we could look right through things, and perhaps it is
fortunate that we have not.  But we can make an instrument which is more sensitive than the human eye and can
get in that a photographic representation of it, and therefore see what we have before called the unseen.  Take it
in sound.  The lowest note on a great organ and about the
lowest note of which the human ear is capable of receiving an impression, is made up of 16 vibrations to the second.  The highest note, which sounds on the human ear
like a very thin attenuated and uncertain sound, has some
30,000 vibrations to the second, but there are vibrations
below 16 and much higher than 30,000, perhaps mounting
into millions.  How do we know what sounds are about
us that we never hear?  A sound once uttered is never lost.
As a pebble dropped into the glassy surface of the pond sets
waves going whose circling vibrations sweep on until they
break upon the shore, so a sound once uttered starts a series
of vibrations which, in the shoreless sea of space, we may
judge, go on forever.  Now, suppose we could make an instrument which was capable of receiving these infinitesimal
sound waves still in the air that were started away back at
the creation, and then could magnify them by some process
which would make them apparent to the human ear, what
should we hear?  The song the morning stars sang together
when the world was made; the mighty grinding of the creation, the loving words of our first parents as they walked
together in the garden, and mother Eve's crooning song
over her first born that came as a comfort for lost paradise?
We should hear sounds of peace and then of battles, the
struggles of early men with their rude weapons, the song of
Miriam by the Red Sea's parted waves, the shouts of Roman legions, the agonies of Gethsemane, the wail of Mary
at the cross.  We should hear the surges of the great human
sea, the songs of gladness and the moanings of de-
spair.  It would be one constant rolling of sound waves, of

the first cries of the new born child, the last gasp of dying
old age; one great whirling wheel of life and death and joy
and sadness, and so on down to today.

Or, if we could gather the light waves into some instrument and magnify them so that they would become appreci-
able to the human eye, what might we see? There are fixed
stars whose light has been 1,000,000 years reaching this
earth. The light the earth returns has been, of course,
the same period in reaching the distant stars. Suppose we
were placed upon one of these stars. We would see the
earth light coming to us, not as it looks today, but full of
the pictures of 1,000,000 years ago, and there in this great
kinetoscope we might see pictured all the world's progress
from the expulsion of Adam and Eve from the garden and
Noah's ark floating on the vast waste of waters, and see
tribes forming from families, and nations from tribes, see
scenes of war and scenes of peace, and watch the development of civilization.

But suppose beyond that we had an instrument which
could bring to us the sounds immediately present about us,
which we do not hear, bring to our vision the objects with
which the air about us may be filled, and which we do not
see, what harmonics may exist to which our ears are not
attuned; what spirits of light may be all around us to which
our eyes are blind?

How small the limit of our senses! We take a telescope
and look into the sky and see the planets whirl in their
courses and beyond a misty nebulae. A more powerful
telescope and this cloud mist develops itself into stars, and
still beyond another misty belt, and so on, the limit only depending upon the power of the instrument used, we pierce
the deeps above us and yet we do not begin to reach the
bounds of light or hardly pierce the surface of the sky.
And if we take a microscope and look at the world beneath
us we look just as far in that direction and still infinity is
is away beyond us. And now we are beginning to know
that there are forces all about us of which mankind has
been unconscious. What may it lead to? Are all things
mere shadows? May it not show after all that the unseen
only is the real, or, as St. Paul says:

"While we look not at the things which are seen, but at
the things which are not seen; for the things which are seen
are temporal; but the things which are not seen are eternal."
—Times, Watertown, N. Y.

# A COMMON CHARGE AGAINST SPIRIT COMMUNICATIONS.

In his "Defense of Modern Spiritualism," Prof. Alfred Russell Walace explains the grounds of the very common charge that so many of the alleged spirit communications are mere repetitions and recitals of a personal rather than intellectual cast, in a perfectly rational and satisfactory way, and his remarks are of special worth at this time, as they were at the time of their first enunciation. Referring to certain statements on the subject by Prof. Huxley, he quotes a brief extract to this effect: "But supposing the phenomena to be genuine, they do not interest me. If anybody would endow me with the faculty of listening to the chatter of old women and curates at the nearest cathedral town, I should decline the privilege, having better things to do. And if the folk in the spiritual world do not talk more wisely and sensibly than their friends report them to do, I put them in the same category." This pasage, Mr. Wallace describes as having been writen with the caustic satire in which the kind-hearted professor occasionaly indulges. But he adds, it can hardly mean that, if it were proved that men realy continued to live after the death of the body, that fact would not interest him, merely because some of their conversation was not up to the standard.

Many scientific men—Prof. Wallace proceeds to comment—deny the spiritual source of the manifestations, on the ground that real, genuine spirits might reasonably be expected not to indulge in discourse upon the commonplace affairs which often form the body of ordinary spiritual comunications. But surely Prof. Huxley, as a naturalist and philosopher, would not admit this to be a reasonable expectation. Does he not hold the doctrine that there can be no effect, mental or physical, without an adequate cause? and that mental states, faculties and idiosyncrasies, that are the result of gradual development and lifelong—or even ancestral—habit, cannot be suddenly changed by any known or imaginable cause? Ad if, as he would very likely admit, a very large majority of those who daily depart this life are persons whose pleasures are sensual rather than intellectual—whence is to come the transforming power which is suddenly, at the mere throwing off of the physical body, to change these into beings able to appreciate and delight in high and intellectual pursuits? The thing would be a miracle—the greatest of miracles: and surely Prof. Huxley is the last man to contemplate innumerable miracles as part of the order of Nature.—The Religio Philosophical Journal.

## SPIRITUALISM IN THE BIBLE.

No one would guess from what periodical the following references are extracted:

Spiritual Gifts—1 Corinthians xii., xiii., xiv.; Romans xii.

Spiritual Circles—Acts ii.

Dreams—Matthew i.; Genesis xi., xxiii., xl.

Test Mediums, Seers and Prophets—Acts v.; John iv.; 1st Samuel ix., xxviii.; Michah iii. 5, 7; Deuteronomy xviii.

Slate Writing—Exodus xxxii., xxxiv.; Deuteronomy x.

Writing on the Wall—Daniel v.

David a Writing Medium—1 Chronicles xxviii. 11, 19.

Psychology—Acts xiii. 9, 11; Mark viii. 22, 25.

Obsession—1 Samuel xvi. 14, 23; 2 Chronicles xviii.; Acts viii. 7, xix. 15.

Fire—Deuteronomy v.; Exodus iii.; Daniel iii.

Materialization—Luke xxvi.; Acts i., xii.; Genesis xviii., xxxii.; John iv., xx.; Exodus iii.; Ezekiel viii.; 1 Corinthians xii.; Joshua v.; Numbers xxii.; Daniel viii.

Mind Reading—Mark ii. 8, 9; Matthew xii. 25.

Healing—Mark iii., v., vii., viii.; Acts iii., v., viii., xiii., xviii., xix.; John v., xi.; Matthew vii. 15, 17; ix. 31, 34; xii.; 2 Kings iv., v., xii.; Ezekiel ii.; Samuel iii., x., xvi.

Open-eyed Mediums—Numbers xxiv. 1, 4.

Shut-eyed Mediums—Acts ix. 1, 19.

Destroying Mediumship—Acts xii. 16, 19.

Developing Mediums—Matthew x.; Mark i.; Acts ii. 4, 18; viii. 15, 19; xix. 11, 12; Ezekiel ii. 1, 10; 1 Samuel iii. 8, 13; x. 1, 11. Prophecy—Revelation vi.

Trance and Voices—Acts x., xi., xxii.

Trumpet and Voices—Revelation i., iv., v., vi., viii, xviii., xix., xxi. Be spiritual—1 John iv. 1.

These are from the Agnostic Journal. It is the most surprising paper imaginable. A good half of its pages are taken up with theosophy; the religion of complete revelation, and certainty about everything, ventilating itself in a paper whose very raison d'etre is inquiry and suspension of judgment. A good deal of space, too, is occupied very worthily by some interesting letters of Mr. Maitland's, on the relation of reason and intuition, which to quote in fragments, would be to spoil, and which space considerations forbid us to quote entire.

—Borderland.

———

"But Peter, standing up with the eleven, lifted his voice and said unto them:     *   .*   *   *   *"

But this is that which was poken by the prophet Joel:

And it shall come to pass in the last days, saith God, I will pour out of my spirit upon all flesh; and your sons and your daughters shall prophesy, and your young men shall see visions, and your old men shall dream dreams.

And on my servants and on my handmaidens I will pour out, in those days, of my spirit; and they shall prophesy."

The Acts ii. 14, 18.

See Joel ii. 28, 29.

# IN THE SPIRIT LAND.

*Mrs. A. C. Rogers Tells of a Remarkable Experience—*
*Works of Spiritualism—Commuuication Estab-*
*lished with Her Dead Husband—Some-*
*thing About Human Souls After*
*They Have Passed Away.*

[Published in the Daily Inter-Ocean, of Chicago, March 23, 1896.]

Grand Hotel des Anglais, Valescure, St. Raphael, France, Feb. 25.—To the Editor.—It is strange that the world has been so long in satisfactorily answering the question: If a man die shall he live again? All kinds of speculations have been indulged in, and many volumes written to prove him immortal, and yet there seems to be a majority that can have no conception of a life that must survive the rigid, mysterious state called death.

As the world has advanced in what is called science and has striven to find an adequate cause for all human power in the wonderful construction of brain and nerve, and for all the phenomena of external nature in a blind force, this perplexing question has become more and more puzzling.

The double nature of man, as recognized in his primitive state, first gave an intimation of the existence of a personality independent of the apparent fleshly one. The fact that in dreams could be seen, felt and heard all that was possible through the senses gave the idea of those phantoms "which will not down," ghosts. A belief in these has, among those thought to be most highly civilized, generally placed the credulous one upon a plane with savages of the lowest type.

Although Herbert Spencer traces the basis of all our institutions, both civil and religious, to primitive ideas founded upon the ghost theory, he afterwards asserts, in substance, that there were no such beings in existence. His reasoning faculty was quick to discern the flaw in his argument, and to realize that things based upon a falsity must be false in themselves. In this dilemma he aserted that those imaginative people must have had a vague sense of the mysterious about them to have formed such conceptions.

But religious and civil institutions were not based upon indefinite conceptions, but upon specific ideas peculiar to all primitive races, and some universal truth must have furnished the foundation. It may be that the primitive or savage races, being nearer to nature, were most susceptible to intuitive truth and needed a peculiar guidance. They felt that there was a power about them that needed to be propitiated or worshipped, and having no idea of the true God, they offered sacrifices to the shades of their ancestry. All traditions, both sacred and profane, attempting to give accounts of these initial efforts at worship record instances where the assistance was given of voices as well as visions.

It is to be doubted whether these came from the highest order of angels or spirits, for it is rational to conclude that spirits of a somewhat congenial order would be in communication. In a book purporting to have been dictated by spirits the statement was made that the angels that talked with Moses and others of the olden time "were of a very inferior order, compared with those that surrounded and guided Jesus," though they were often mistaken for the God of the universe.

This might explain some of the seeming contradictions and incongruities noticeable in the character of the messages given in the respective epochs. Sacrifices and burnt offerings commanded by angels to be offered to the Lord in the days of Moses were spoken of as abominations in the sight of the same Lord when Christ came with a new message to the world. Every age and religion seem to have received the inspiration for which they were fitted, and none has been left to grope in complete darkness. All beginnings, in religion as in everything else, have been crude, and must have been so in order to the progress required by a universal law of nature or of God.

Notwithstanding the many instances recorded of spirit influence and communion, the testimony of seers of undoubted intelligence and unimpeachable character, such as Swedenborg, and of investigations in the interest of science

bv psychical societies that have collected many facts cor-
roborative of the existence of a life and a body independent
of the grosser one, apparent to the senses, incredulity is
the rule.   Under these circumstances, I feel it incumbent
upon me to add my feeble testimony by giving to the public
a condensed account of an experience in many respects remarkable.

I do this alone in the interest of truth, and because I feel
that I should be held accountable for keeping to myself
what I consider knowledge that must have been given me
for some other purpose than my own gratification. Perhaps
I should at an earlier date have made this disclosure public, but in consideration of family and friends, as well as my
own interest, I have refrained from so doing.   I have never
intended to carry this mystery in concealment to the grave,
but, having reached the age allotted to man on this earth,
feel that I may not have a great while longer for delibera-
ation, and write while I may.

I feel sure that my paper will be read with incredulity by
most.   Generally it will be thought that I have been deceived in some way, and a lower estimate will be placed
upon my judgment and discriminating faculties.   Many will
account for the unusual phenomena by telepathy, mind
reading, and perhaps some phase of hypnotism, these forms
of mind influence having received much attention of late.
1 am superficially acquainted with some of the possibilities
claimed for the theories bearing the above names, and admit that some of the phenomena might have been produced by mind reading or telepathy; but that all, or most
of them, covering so long a period of time were the result of
these subtle powers I cannot for a moment admit.

Responisbility rests alone upon myself, and let the result be as it may, I must discharge what I consider to be
my duty.

Shortly after the sudden death of my husband, Judge
John G. Rogers, in 1887, a spiritualist came to me with
what purported to be a message from him to me.   He was
not an acquaintance of mine, but had known my husband,
and after seeing and talking with him I could not doubt
his sincerity, and, although thinking he might be deceived
in some way, decided to investigate for myself.   Of the
medium I had never heard, nor had she of myself, and no
one knew of the time selected for my first visit.

I found the medium in bed, and in addition to habitual invalidism she was suffering from a hurt received in some accident. She said she doubted whether she could, in her state, give me the least satisfaction; but in a short time, whilst talking upon indifferent subjects, her eyes gradually closed, and she seemed to be in an unconscious state.

Then her voice and manner changed completely, and her body seemed to be the organ of a man's speech. He claimed to be in control of the medium, and said he knew what had influenced me to make the visit—that my husband was standing near, but was so full of emotion that he doubted whether he could speak to me, but the attempt would be made. In a short while her manner and voice again changed, and she seemed to be controlled by one laboring under deep emotion, who gave the name of my husband.

He said, in a halting voice, that after realizing that he was in another world, he felt he must have some word with his family, and, although not believing in spiritualism, sought every means of communicating, and that in his perplexity a wife of the spiritualist who had brought me the message told of her intercourse with her husband on the earth, and offered him assistance, saying that at first it would be difficult, but by practice would become easy.

The manner of the medium, and the nature of the communication, assured me that it was really my husband who was talking with me.

We mentioned facts known only to ourselves, and in many ways that it would not be possible for me to explain convinced me that I had communicated with one thought to be dead. He begged me not to think of him in this light, but as being nearer to me, and understanding me better than ever before. He said the value of my visit to him could not be estimated in any way, as we consider values, but was worth more than all the world to him, and that I must continue to come until he could talk with me with less difficulty.

Under the circumstances I could not but comply with his request, and have for the past eight years or so gone at stated times to visit a beloved one for his pleasure, and as he feels, of profit, as well as my own. I have not, however, gone often enough to lose any of my interest in my present life, feeling much more than before my experience the obligations resting upon me in a world so near and in such close relationship to the one to which we must soon go.

I have had communication with many others who have convinced me of their identity. Parents, grandparents, other relatives, and many friends, also children that passed away in infancy, have been described to me in their present state of maturity. For all, the first communications seemed difficult, often given in whispers.

The medium, although not a woman we would call cultured, impressed me as one of honesty and refinement. Of her sincerity and purity of intention I could not doubt, and the opinion first formed has been strengthened by the acquaintance of years. She said she had possesed the power to see and communicate with spirits since her earliest recollection, and that they seemed to her as real and not any more mysterious than human beings.

Although showing a kindly and friendly disposition, she seemed to have no curiosity in regard to my connections and surroundings, and said she would prefer not to know of them. It was two years before she knew my name. She has two controls, a man and a woman, the first claiming to have passed away from this life about 100 years ago, and the other to have been in that world for as many as 600 years.

They said this was the work they had been called upon to do, and that so long as the medium should live they would strive to do all the good possible through her, and then take up some other work. It might be they would not be called upon to visit the earth again, though it would always be possible. It seems natural that the familiar path back to the earth would be plainer and more easily traversed than any leading to unknown heights.

The woman control says she is more fortunate than many, in being able to go both high and low, that she can witness the delights of those who have attained to blessedness through tribulation, and the griefs of those who are still struggling in sorrow and remorse.

Some of the things told me of the next life to this were opposed to my preconceived ideas, and were not always agreeable; but it was truth I was seeking and not to have my own opinions substantiated. It had been a source of pleasure to feel that with the change called death came perfect rest and satisfaction, though upon reflection this is opposed to reason and the general order of nature, and I was told that often the bitterest experiences came and a period of darkness ensued of long or short duration according to circumstances.

Remorse was felt for a misspent life, for neglect of duties, and sorrow for even the mistakes of life; that it was only through striving for the highest and best that light appeared and happiness dawned. The character and peculiarities of each were transferred to a different plane, and that these decided the position and surroundings in the new life. If the affections were strongly bound to persons or things of the earth spirits would for a greater or less time hover about them, and in the case of the former strive to influence and bless.

The desire for communion with the earth's inhabitants was almost universal, and in many cases would result in mutual good; that is, where the best conditions existed.

The life there is so much more real than this, as to cast over it a shadow, making it seem insignificant.

I have been told of the improvement and development of some of my friends, and of their hopes for further advancement.

If I had needed any confirmation of the reality of these communications it has been given in the case of a very few friends that have passed to the next life since I have had my peculiar experience, and with whom I had talked about it. Of course the medium could have known nothing of this.

In one case apologies were made for the incredulity with which my narrative had been received, and it was said some benefit had been derived from it, notwithstanding the doubt with which it had been received; that a true knowledge of the conditions in the next life greatly added in progress and development there.

The general law of progress was said to hold there as well as here, and we may reasonably conclude that it continues to eternity, if we can conceive of it. It would be futile to conjecture as to what is possible to the spirit in the ages before it. In the language of St. Paul, "It doth not yet appear what we shall be."

As progress takes place slowly in the natural world, and we find no sudden breaks or transitions, we must conclude that the plane of life immediately succeeding the present one must partake in some measure of the characteristics of this, that it is the reality of which this is the shadow.

We could not desire a state of existence of which we could form no conception for the want of a type or symbol.

We very much mistake true conditions when we ascribe to spirits perfect knowledge. They often say: "We do not know everything." There are the same differences of opinion among them, though they have advantages over the inhabitants of this sphere in increased opportunities for advancement, when relieved of the trammeling influence of the flesh. Aristotle is said to have thought that the reasoning faculty could never do perfect work until the attainment of this condition.

There must be in the next sphere spirits that adhere for a longer or shorter period to their peculiar religious views. I can give a case in illustration: When a young girl I was frequently at the house of an aunt, who was considered unusually pious. I do not remember exactly when she died, though it must have been thirty-five or forty years ago. She had not consciously entered my mind for many years. At one time, while having a sitting, a house was minutely described to me; I failed to recognize it.

Then it was said a large, gray cat lies on the rug before the fire, and the figure and face of this aunt were so graphically given, with the dress that was so peculiar to her, that she was plainly brought to view with her surroundings. It was then said that she was just beginning to see the light, and had progressed so slowly, because of erroneous views pertinaciously held. She had been looking all these years for her Savior.

Another relative of mine, who had been a pronounced agnostic, died at his home in St. Louis. Although not thinking of him, and without expecting a communication, of course, he controlled the medium, and held out trembling hands, and, in a whisper, gave the name by which I had been accustomed to call him—a very peculiar name. one I am sure the medium had never in her life heard.

Before this demonstration was made, however, I was told a spirit wished to talk with me, whose name was James, as given by my mother. A confirmatory coincidence in this case is that I had never called him by the name familiar to my mother, and, in speaking to me, the one by which I had called him was given. He said in a very feeble manner that he would try some other time to talk with me. As he never appeared again, I asked the control why it was. She replied: "He was not prepared for this world and is like one in prison."

Not very long after, and without having heard of my experience, a brother in the South wrote that while one of his children was experimenting with a contrivance for telling fortunes—I forget its name, though it is something in the nature of a planchette—the name of this relative was distinctly written and after it: "I am in prison."

There are, of course, many personal details I might give, but doubt whether they would make more probable the truth of what I am trying to enforce, the reality of a life after this, and the possibility of its being in communion with this under favorable conditions.

Certainly they would not impress others as they did me, and would not be considered of scientific value. The evidence to me has amounted to demonstration, and I should consider myself as deficient in reasoning powers were I to reject it as worthless, or even not entirely convincing. I know that all who have attempted investigations in the same direction have not been so fortunate, and in many instances have been sadly disappointed.

All mediums are not honest, and, besides, it seems to be necessary that there should be a kind of adaptability between the medium and the sitter. What that is it would be impossible to say. The medium through whom I talked had said that for some she could get nothing satisfactory and that the effect of some investigators upon herself was rather distressing, they having a tendency to cause great exhaustion. She could not account for this fact herself.

I shall now try to answer a question that will be asked by many. What good could come from spirit communion, even if general? If no other good should ensue than settling the question of the immortality of the soul its advantage could not be computed. But besides this, the happiness of lives upon the earth might be greatly enhanced. Although the change that takes place at what is called death is not so great as is generally imagined, it must bring its advantages.

Emerson has said: "Whatever is natural and universal

must be beneficent." The flesh must be a clog to the free exercise of the powers of the soul, and when freed from it they must expand inconceivably in every direction. The natural effect of every cause must be much more apparent, giving an insight into the future impossible to us. This might be in a measure imparted as a warning against certain tendencies of thought and courses of action.

I should like to add here that I have been told by spirits that nothing that would have a deleterious effect upon the inhabitants of the earth would be permitted utterance.

The approach of the inevitable change called death would no longer be contemplated with horror and dismay; but welcomed as a friend, indispensible to conditions more to be desired than any to be feared on the earth sphere.

What I have said will lead to another question more difficult to answer, which is: If all these advantages would result from spirit intercourse, why has it not been vouchsafed to the inhabitants of the earth universally? Man's own departure from the true path is the only reason I can give. I am of the opinion that many intuitive truths and spiritual gifts have been lost to the world through increasing interest in the things of sense, and that research in this direction for hidden and absolute truth always has been and ever will be disastrous. Our scientists must make a new departure before they can reach convincing and satisfactory truth. They must learn that there are many more far-reaching facts to be discovered in the subjective than in the objective, and that both paths must be traversed.

With a general recognition of the fact of the nearness and interest of the spirit world, what solace would come at the hour of death. As the grasp of the hands of our earthly friends must be relinquished at the last hour, welcoming and loving ones would be extended to guide us to more enduring and more blessed homes beyond.

—A. C. Rogers.

# SURVIVAL AFTER DEATH.

## Mr. Stead Interviewed.

Mr. Stead has been interviewed in Chicago for the Riligio-Philosophical Journal, of that city, and the following is taken from the article which appears in the issue for December 23rd:—

## A PERPLEXING QUESTION.

We began by asking Mr. Stead what, in his opinion, is the strongest and most convincing proof of the mind's survival of the body. Mr. Stead hesitated some time before giving an answer, and then said that he did not think he was competent to decide which was the strongest proof, but, so far as he had gone, he thought that the strongest proof in favor of human personality was supplied by the evidences that are daily multiplying of human intelligences communicating while still in the body with other persons through agencies which are independent of the body. "That is, it seems," said Mr. Stead, "that if my personality is so much greater than, and more complex than the fragment of it, of which alone I am conscious, and if I am able to prove the existence and functioning of my own personality, independently of my conscious mind, or of the senses of the body, it seems rational to think that this mind, which operates independently of the body, will continue existing after the body has dissolved. So long as it was possible to imagine that human intelligence was simply a product of the brain which dissolved at death, the proof seemed the other way, but when you are confronted at every turn with evidence that a man's personality can function independently of his body, you are no more inclined to believe that you cease to exist when your body dies than that you pass out of existence when you lay down the telephone and ring it off."

We then asked, "What class of so-called spirit manifestations seems to you to possess the greatest evidential value?" "Those manifestations," said Mr. Stead, "which supply the greatest amount of evidence. The ghost is a very fitful creature, and the evidence which we gather from his appearance is small compared with the value of the evidence obtained from manifestations which are more under control. So far as I have seen, materializing seances do not amount to much. I do not say they may not have very high value some times, but, so far as I am concerned, I have not seen anything that was at all conclusive in the way of materialization, which is no doubt my misfortune if not my fault. But I think the evidence obtained by automatic writing or by trance mediumship gives more conclusive evidences than what is obtained in any other way."

## IS THERE ANY POSITIVE PROOF?

"But, Mr. Stead," we asked, "do you think that there is positive proof that the spirits of the departed manifest themselves to those in the flesh?" "That is a subject," he said, "upon which I do not wish to dogmatise. Positive proof means, I suppose, proof as clear as that which demonstrates a problem in arithmetic. I think that Mr. Miers, when he was over in Chicago, put the truth of the matter very well when he declared that he believed that when all deductions had been made in the shape of coincidences, telepathy, and other causes, there remains an irreducible minimum of evidence in favor of the hypothesis of the spirit's returning to earth, which could hardly be explained away. I am not quoting his words, but I think that was the gist of what he said."

"Do you think that there is proof positive that those who manifest themselves are the identical persons that they profess to be?" Again Mr. Stead objected to the words proof positive. "All I can say is that there is a reasonable degree of certainty on the subject. Of course, if Mr. Hudson's theory be correct, and the unconscious mind is as omniscient as God Almighty, and is absolutely without any moral sense, then such proof is absolutely impossible, because, according to Mr. Hudson's theory, it is easy for your own unconscious mind to possess itself of all the information necessary to furnish the most conclusive evidence, and at the same time to deceive you by asserting that the information was communicated by some dead friend. But Mr. Hudson's theory seems to me to be much more incredible than the alternative. I have proof positive just as much as the cashier of a bank has when he believes the paper was drawn by the man who signed it. Bank cashiers seldom have proof in a legal sense, but they have reasonable assurance, and that I think we may claim to have in spirit communications."

"To revert to the automatic writing, Mr. Stead. Have you had any experiences since you came to this country?"

"Yes; but I have been so very busy that I have not had time to conduct such experiments, and besides, when you have to wait twenty days for confirmation of these experiences it is rather a bar. At the same time I find that distance makes no difference, and I get messages from my friends across the Atlantic as readily as from across the street."

"Do you think that communication with the so-called dead should be cultivated?"

"It depends upon whether or not you consider the so-called dead desirable acquaintances. There are many of my acquaintances with whom I do not desire to continue to communicate with one moment longer than I can help, but if you love anybody it seems quite as unnatural that you should cease to wish to communicate with him because he has put off his body as because he had bought a new pair of boots."

## IS MEDIUMSHIP HARMFUL?

"Do you think mediumship is generally detrimental to the medium?"

"My experience is not wide enough, but among my personal friends who have mediumistic gifts I do not know of one who suffers from the exercise of them, nor do I know of one who would give it up even if it did entail a certain amount of physical exhaustion. I do know of cases where mediums have lost control of themselves. I have also known of cases in which mediums have exhausted themselves by excessive mediumship, and have been brought into a state of prostration from which they have sought to emerge by means of drinking. It is somewhat risky business, being a medium, although I suppose I am a medium in a kind of a way. I do not consider it

dangerous being a medium as long as you have control
of yourself, and I have always been master of my own
hand, although I may allow it now and then to be used by
another mind than my own."

"Have you discovered much fraud practiced under the
name of mediumship?"

"Not so much fraud as folly. However, the flap-doodle
some mortals can swallow when it is vamped up with
spiritualistic dressing is almost inconceivable. At the same
time there are fraudulent mediums, and it is one of the mis-
fortunes of the regular practice of mediumship for pay that
there is a constant temptation to forge a communication
when no genuine message can be secured."

"Can you state some of the most essential, physical and
intellectual conditions favorable to mediumship?"

"Health, I should say, and the ability to place your mind
in a passive condition. Mr. Stainton Moses always believed that it would be useless for me to hope to obtain
any manifestations or make any progress in manifestations from the unseen world, for he said my mind was too
full of the vibrations of intense mental activity, and my
whole life was lived in an unending whirl."

## THE BIBLE AND MODERN SPIRITUALISM.

"Do you find in the Bible confirmation of modern
Spiritualism?"

The answer was, "I think you put the cart before the
horse. You can find in modern Spiritualism confirmation
enough and to spare of the Bible. In Borderland I have
begun a series of papers touching this subject. I began
the series with the Prophet Elijah. Almost all the phenomena that I find in the Bible are being reproduced, as
anyone will find if he will take the trouble to inquire. The
gift of hearing, seeing, levitation, the gift of premonition,
the power to hold communication with good and evil
spirits, and to communicate with those who have passed
into the invisible world—all these are in the Bible, and,
as everyone who knows anything of the subject will admit,
they are of more or less frequent occurrence in the world
today."

"What have you to say as to the Satanic theory of
Spiritualism?"

"It is a very natural theory. Everything used to be credited to the devil that people did not understand—thunder-storms, earthquakes, any phenomena that were out of the ordinary run, if they could not be credited to some divine power, were put down to the author of all evil. So it is now as to these communications from the other side. If we are to try the spirits by the same rules of common sense that we try communications received from human beings who are still in the body, we should find that some are good as good can be while others are bad, while the great mass of them talk drivel and flap-doodle. There is nothing satanic about it, so far as I can see, unless the influence of an evil intelligence, whether in the body or out of the body, can be said to be satanic "—Borderland, January, 1894.

# VISIBLE AND INVISIBLE.

Only the grosser forms of matter are visible to the eye. We cannot see an atom, a molecule, or electricity. The most powerful material agents are invisible. We behold a few acres of the earth; we perceive rocks and minerals; also vegetable forms, from the tiniest flowers to the great trees that have braved the storms of centuries; animal growths, from the microscopic insect to the huge mastodon; the illimitable universe, with no conceivable center or circumference, boundless, infinite; and finally man, endowed with mind, soul, and a radiant spark of the divine spirit; we recognize life, consciousness, substance, and rest in the conviction that the universe, visible and invisible, is the thought of the Divine Mind in expression. Consciousness is everywhere, life everywhere, substance everywhere. Manifestations of Supreme Power are discernible in intelligence, force and matter; superior to their manifestations is the eternal and unknowable; but visible Nature, which we can study and know, is the outward expression or manifestation of the invisible, the infinite and the eternal. There we find, as Pope says,

"That God of nature who within us still
Inclines our actions, not constrains our will."

By the close study of nature we can pass from the visible to the invisible, as by a ladder reaching from earth to heaven. . Note the steps as indicated by a modern scientist: "We pass from solid matters, such as metals, to the liquids; from the liquids to the gases; from the gases to radiant matter; from radiant matter to the forces of nature—gravitation, magnetism, light; from force to sensation; from sensation to thought, idea, purpose! Here, too, as with animal and vegetable life, we may well believe, that there is no break in continuity." How soon we pass from the visible to the invisible. In fact we begin with the invisible. A material atom no one can see. It is the indivisible and infinitesimal form of all matter. Science can weigh atoms, but not a single atom. Molecules, made up of atoms, are also invisible, yet science tells us how many there are in a given space, how heavy they are, and how swiftly they move about. A mass, or object, the smallest visible form of matter, is formed of molecules, each one of which, it is said, is "about as much smaller than a pea as an orange is smaller than our earth." We fail to comprehend the infinitely little or the infinitely great.

The more attenuated the forms of matter the farther apart molecules are, and yet they never touch each other in rock or metal. When many molecules are massed in a single form matter is visible. Invisible forms of matter are most powerful. Steam, in which molecules are farther apart than in water, is more powerful than water. Expansion develops power. Electricity more subtile than steam, is vastly more powerful. Yet, without mind to control and direct it, electricity would be today an untamed force and economically valueless. Even matter cannot act on matter without the intervention of mind. The hammer does not drive the nail into the board and fasten it to a building. Nor is it the man's arm and head that drive it. Place nail and hammer and board together; attach a dead man's hand to the hammer, with head and body perfect; apply steam or electricity; the nail will never be driven. A higher force is needed—the force of life itself, acting through the machinery of brain, body, arm and hammer, all directed by intelligence. The man himself, by virtue of invisible power within, drives the nail. Spenser, earliest of English poets, said:

> "For of the soul the body form doth take,
> For soul is form and doth the body make."

The visible is a small part of human life. For the child, that unconsciously breathes a few times and passes away, as for the centenarian who, equally unconscious from the infirmities of age, leaves a worn-out tenement, this life is short. Briefest or brief rest assured the earthly incarnation is essential to the evolution of the soul. Not for fame, wealth, glory, do the saviors and benefactors of the race perform their mission. They are apostles of the Divine Unseen. Serve humanity and live. Serve self and die. Such is the eternal law. This visible life is a strange compound of weakness and strength. Deeply implanted is the sense of immortality, and yet it is strangely set aside for the baubles of space and time. Immortal aspirations and desires alone can loosen the shackles that bind the soul to earth. Material force is a visible manifestation of the spiritual reality behind or within it. Separate from the life there is no force. He who accepts the material and denies the spiritual admits the effect but ignores the cause. Accepting both the spiritual and the material, who shall set a limit to the manifestations of the former through the latter? Paul's words come to mind: "For the invisible things of him from the creation of the world are clearly seen, being understood by the things that are made."

Inspirations come to prophets, poets and seers from the realms invisible to mortal eyes. Martyrs welcome torture and death sustained by a mental power that renders nerves insensible to pain. Desire for earthly honor never has such a sublime effect. Marvelous is this occult power. Note the experience of Paul with the viper fastened to his hand, and the poison having no effect for "he shook off the beast into the fire and felt no harm." Here was manifested the supremacy of mind. It counteracted the effect of the poison of the serpent's fangs. So, too, the invisible is the factor in every life. It is no longer regarded as evidence of intellectual weakness or delusion to accept the teachings of Christ, the Great Healer, and of his apostles, as facts that may be repeated; or to believe that angels walk by our side in crowded cities and lonely woodlands, and watch us when we sleep, though the human eye then sees nothing, not even material objects, for it can no more discern spiritual presences than it can, without a microscope, see the multitude of living creatures in a drop of dew or a ray of sunlight.

Scenes continually change around us; we note the impermanency of all material forms; to get the most of life we too must change to a new and higher environment: but here we should live near to Nature, calmly passing from one day to another through nights of rest and without sorrow for the dead past; live aright in the present and fear not the issues of the future. The invisible furnishes a lofty ideal; there is no work too humble or too great for men to do; no unselfish aim too high for human endeavor: and, as the cycles run their course, the highest welfare of humanity is attained by the faithful performance of present duties while striving for the loftiest ideal. Thereby the visible and invisible are united in achieving for the human race the greatest good both now and hereafter.

> "On wings of deeds the soul must mount!
> When God shall call us, from afar,
> Ourselves, and not our words, will count—
> Not what we said, but what we are."

—Jackson. (Mich.) Patriot.
Republished in Religio-Philosophical Journal.

# IS SPIRITUALISM A RELIGION?

### The Fundamental Principles Clearly Outlined by Rev. A. J. Weaver.

Is Spiritualism a religion? That depends upon what religion is. If it consists of certain rites, ceremonies and forms of worship; of going to church, and observing the Sabbath as the Lord's day; of offering praise and prayer to a personal, outside God, above nature, up in heaven, from whom all blessings flow; or if it consists in believing that Jesus was the Messiah, the only begotten son of God, and that by no other name can we be saved; or that the Bible is the word of God, and contains all we ever need to know concernig immortal truth, and is a sufficient rule of faith and practice; or if religion is simply faith and trust in the future, in God, in the Bible, and the church, whether sanctioned by reason or not; if religion consists in accepting Christianity or any other of the great religions of the world as unquestioned truth, then Spiritualism is not a religion, nor any part of religion.

But if religion is the recognized feeling of obligation we are under to righteousness, goodness, and truth; if religion is the reverence we feel for the infinite life, power and intelligence which pervades the universe; if religion is the desire to commune with heaven, and to rise daily into higher states of thought, of feeling, and of life, then most emphatically do I say that Spiritualism is a religion.

What is Spiritualism? What is the one thing that differentiates it from every other phase of religious thought? It is the recognized fact that the inhabitants of the immortal world, who once were clothed in flesh, can come and identify themselves to us under certain conditions. By this knowledge, and what grows out of it, we are separated from all other bodies of religious people.

But, aside from this, we hold, in common with all the Christian sects, a firm faith in all the graces and virtues of a progressive, spiritual life. We trust implicitly in the goodness, and have faith in the power of the Infinite, whose almighty forces in nature convince us of our weakness and dependence. We realize we are but a speck in the Infinite presence, in which we are bathed. We acknowledge our utter dependence upon a power which entirely transcends our comprehension. We know we are weak and crude, but we aspire for wisdom and growth in the verities of truth and goodness.

Like other religious people, we accept with our whole hearts every truth contained in the Bible. We recognize and proclaim every noble quality exhibited in the person and life of Jesus. We acknowledge our obligation to embody in thought and act every principle presented by Him which will add to our spiritual growth.

We do not claim that Spiritualism is the one perfect religion, or possesses the whole of religious truth, or is without error. But we do claim it contains more truth with less error than any form of religion on earth. We admit that Christianity, in all its forms, even evangelical, embodies, in a greater or less degree, some of the enduring qualities of a true life. Spiritualism accepts and uses all these while it rejects its errors which, on its very face, are absurd and in violation of nature.

Unitarianism is the best and highest form of Christianity, and we accept it. But we are larger than Unitarianism. We embrace it, but we rise higher, and take in other truths of the greatest importance.

If Unitarians are a religious body, much more are Spiritualists, because we are not only Unitarians, but something more.

If Universalists are a religious body, much more are we, because we are not only believers in the ultimate upward growth of all human beings in this world or the next, or both, but we accept an additional truth to which the Universalist body has not yet attained.

I was ordained, and preached for many years in the Universalist ministry. When I stepped into the Spiritualist ranks I did not lay aside or leave behind any of my religion, but added more religious truth to what I already had. If what I believed was acknowledged by the law of the land as a religion, much more is my belief now a religion, on the principle that the greater includes the less.

If the truths of Christianity are a religion, then Spiritual-

ism is a religion, for we accept every truth of Christianity.
I repeat every truth of Christianity, and we have added to
and enlarged that truth by the modern revelation, which
is but a rebirth, in a more perfect form, of the truth lost
amid the dogmatic strife of early Christianity: the stone
the builders rejected.

It is claimed by some that Spiritualism is a science. It it
a better statement to say it is a religion based on science, as
every religion should be. By this I do not deny the need of
faith. I do not deny there should be a place for faith in
every religious system. In my own thought and life I
give it place and recognize its use. The principle on which
I act is to obtain knowledge on every line of thought so
far as I am able, and then to trust unhesitatingly in the Infinite that all will be well.                          -

It is not Spiritualists who are infidels; who doubt the
everlasting integrity, uprightness and unchangeableness
of Infinity as manifested in the existing order of nature. It
is Christians who doubt this. It is Christians who claim
there is to be a smash up of the world, a destruction of
her forces, and a suspension of her laws. It is Christians
who regard the Infinite as unstable, having laws and yet
finding it needful at times to suspend them, to confer
some special benefit; having certain laws for this world,
but setting them aside in the world to come.

Spiritualists, far above orthodox Christians, believe the
universe is built on goodness, and is governed by wisdom.
They develop self-reliance by depending on self and not on
God so far as they have strength, but when they get beyond their depth, and mystery confronts them, they put
implicit confidence in the Infinite and unseen. They do not
pretend to know much about the almighty Power that un-
derlies the universe of matter and mind, but they sink into
the arms of slumber every night with an abiding faith that,
whether they awaken in this world or the next, whether
they meet joy or suffering, success or disaster, all is well.
They do not feel the Infinite need be beseeched to give to
them protective care; like Whittier, they believe they cannot wander away from the Infinite presence and protection.
They rest on their own knowledge, so far as they are able,
because that develops inward strength; beyond that they
do not tremble and make life miserable by "fear of God,"
but, like a child, trust in the wisdom and power of the un-

seen, and by that sublime trust live in abiding peace.

I believe it is debasing to shrink with fear before the stupendous mysteries of existence. It is true that more or less "we walk by faith." Knowledge is first to be desired, but while we are seeking for knowledge we trust and wait. We cross the threshold of knowledge with uncovered head and shoeless feet. We are in the presence of the Infinite. We are small and weak, but the Infinite is boundless and strong. We are ignorant, but wisdom fills the universe. We are short-sighted, but, with the almighty forces beneath and above us, we can trust.

It may be said, and it is said, that Spiritualists have no religion because they reject so much of the Christian religion. But we reject nothing only what is false. We deny that Jesus is our savior; that he is the only begotten son of God; that he was the Jewish Messiah; that he was anything but a man; but the Unitarians do all this, and they are a religious body.

We deny the Bible to be the "word of God," but the higher critics do the same, and many of them are prominent Christians and workers in the Christian church.

The Jews go much further than some Spiritualists in what they reject. They even reject Christ, the whole New Testament, and Christianity itself as a divine revelation, and yet the law recognizes them as being a religious body.

It is utterly inconsistent to deny that Spiritualists are also a religious body. They differ from the Unitarians only in the fact that, instead of accepting a future life as a hope or faith, they accept it as positive knowledge, based on actual demonstration. Instead of being a weakness, this should be regarded as a strength.

To be sure the Spiritualists are cut aloof from all other religious people in that they believe in the reality of the psychic phenomena; but the early Christians believed in them, Jesus and his disciples produced them, the primitive church abounded with them, and the Bible is a visible record of them.

On what ground can we be denied the legal rights and protection which other religious bodies enjoy? Not on the ground of what we believe nor on the ground of what we reject. I can find no ground except that of absolute prejudice. Such denial is unjust, inconsistent, bigoted and beneath the dignity of well-balanced minds.

The time will come when civilization will look back with shame and disgust upon its treatment of Spiritualism. The pages of history which this age is writing in its treatment of Spiritualism will, in the future, be regretted as we now regret the opposition the world brought to bear against Christ and early Christianity, against the discoveries of science, and against the abolition of slavery and the drink habit.—Light of Truth.

# SPIRITUALISM.

## *It Is Rampant in Paris.*

A recent number of The Literary Digest contains the following:

That an excess of skepticism or unbelief always brings a reaction toward superstition is a well-known fact in the history of religions. This reaction is now being experienced in France, which has acquired a reputation for being the most irreligious of all countries. If we are to believe Jules Bois, who writes a long article on the subject of "Miracles at Paris" to Le Figaro (October 12th) that country is now on the return swing of the pendulum, which is just at present bearing her through a spasmodic interest in Spiritualism. Says M. Bois:

"We must say this much in justice to Spiritualism: It has been the first to raise the standard of revolt against the materialism in which we are wallowing. M. Zola has, perhaps, created the symbolist school by the excess of his naturalism. Spiritualism is a much deeper reaction against the atheism of Proudhon, the skepticism of Renan, the braggings of Buchner. I know that crazy people have been mixed up in it, but there are weak heads everywhere. In fact, it has been the consolation and pleasure of the highest minds.

"Mme. de Girardin passed the last years of her life in the company of Mme. de Levigne, of Sappho, of Moliere, of Sedaine, of Shakespeare.

"Auguste Vacquerie, in his Miettes de l'Historie, relates that at Jersey he made the tables talk on the shores of the sea. 'I believe in spirits as firmly as I do in donkeys,' he affirmed. For him, the scale of beings reached from man to the sky, as from man to the abysses of the earth. * * * Victorien Sardou, thanks to the spirits, amused himself with making little palaces on paper with musical notes. Flammarion renewed the science of the heavens with these studies. M. Jules Lormina refreshed his imagination with them, and M. Gilbert Augustin-Thierny, in many romances, exalts reincarnation, that Spiritualistic dogma.

"In our days the movement has grown in innumerable directions. The painters, usually so material, have set to work to reproduce the miracles. M. Odilon Redon in his lithographs, recreates the terrors of the wandering ghosts. M. James Tissot puts his talent at the service of the 'materializations' of phantoms. Count Antoine de La Rochefoucauld, yet more subtle, seizes the angelic soul at the moment when it leaves the body, in the state of ecstacy. M. Vatere Bernard draws harpies; M. Phillippe—Charles Blache surprises the melancholy spirit at the threshold of the invisible; M. Henry de Malvost invokes the devil himself with his pencil. * * * The celebrated musician, Mlle. Auguste Holmes, receives messages from the beyond; the poetess Mme. Zola-Dorian hears the voices of the invisible.

"What shall I say? The boulevard itself forgets to rail, or rather dares not. On the Tortoric terrace M. Aurelien Scholl relates to me the prodigies of Home, who altered the hour on a clock without touching it, and Maurice Montigut still shivers at the recollection of his juvenile experience at table-turning.

"M. Paul Adam has suffered for more than a year from the assault of a ghost, who gives him troublesome advice. At the house of the Baroness Deslandes we see spirits writing and rapping. * * * The modern chiefs of the state have it appears ,the same love of miracles as the emperors and kings of the Middle Ages, who lived in the company of astrologers sorcerers and alchemists. The correspondent of the Daily News having asked of President Carnot his religious belief, the latter answered that he was a disciple of Allan Kardec, but that he adhered to the Catholic religion for State reasons. And every one knows of the tears shed by Queen Victoria over the death of the medium who had given her the opportunity of talking with the Prince Consort."

After filling a couple of columns with stories of Parisian ghosts, mediums, table-turnings and rappings, all in the good old style, M. Bois closes with the following reflections:

"Unfortunately, the majority of the spirits are too simple: sometimes they are even ignorant and superstitious. On how many of their communications do the asses' ears of King Midas appear? One of their apostles, who is possessed of a wise and inspired intellect. M. Bouvery, confesses to me that in certain seances they go so far as to punish the spirits. Spiritualism, to be born anew, must undergo the ordeal of the phenix. Today, rebaptized in America as 'the new Spiritualism,' disembarrassed of its old errors, it attempts, in the hands of savants such as William Crookes, Aksakoff, Richet, De Rochas, Gibier, Baraduc and Dariex, to furnish experimental proof of the survival of the ego. If the soul survives, what a source of resignation for the suffering, what a balm for the wounds of society! I know of no generous intellect capable of a lack of interest in so great an undertaking."

A few years ago journals like the Literary Digest would have considered it a disgrace and ruinous to admit anything to its columns bearing favorably or unfavorably upon the subject of Spiritualism.

If an unbeliever or (non-knower) in the subject, Jules Bois is at least fair in his treatment of it. The fact is Spiritualism is permeating the current literature of the day in all languages. It is forcing its way into the churches and their pulpits. It is in the air everywhere, and it is only a question of time, and not very long, when it will be universally recognized, acknowledged and accepted as true. If it does have its shades, it has also its most glorious, brilliant lights, the latter more than overwhelming the former.                          H. V. SWERINGEN.

Published in Progressive Thinker, Jan. 25, 1896.

# APPENDIX.

Note 1.—

John Clark Ridpath, LL. D., in an instructive article, entitled "The Bond and the Dollar," published in the "Arena" for January, February and March, 1896, advances some ideas similar to some expressed by Mr. Blaine. In an extensive note under this article, on pages 271 and 272, of the January number, Mr. Ridpath gives the debt of the United States, on March 1, 1866, the debt of the United States in November, 1895, the average prices current on March 1st, 1866, of wheat, flour, cotton, mess pork, sugar, wool, beef, bar iron and farming lands in Ohio and Mississippi valleys (approximately) and the average prices current of the same on November 10th, 1895. In this note it is shown that the national debt on March 1, 1866, would purchase 1,486,842,105 bushels of wheat and the national debt at the close of 1895 would purchase 2,133,620,689 bushels of wheat, an excess of 646,778,584 bushels in favor of the debt of 1895; that the debt of '66 would purchase 262,790,-697 barrels of flour, and the debt of '95, 353,571,428 barrels of flour, an excess in favor of the debt of 1895 of 90,780,731 barrels; that the debt of '66 would purchese 5,885,416,666 pounds of cotton and the debt of '95, 14,558,823,529 pounds of cotton, an excess in favor of the debt of 1895 of 8,673,406,863 pounds; that the debt of '66 would purchase 99,576,313 barrels of mess pork and the debt of '95 150,915,853 barrels of mess pork, an excess in favor of the debt of '95 of 51,339,540 barrels; that the debt of '66 would purchase 5,330,188,679 pounds of wool and the debt of '95, 5,755,813,953 pounds of wool, an excess in favor of the debt of '95 of 425,625,274 pounds; that the debt of '66 would purchase 41,851,851,851 pounds of bar iron, and the debt of '95 46,348,314,606 pounds of bar iron, an excess in favor of the debt of '95 of 4,496,462,755; that the debt of '66 would purchase 25,393,348,314 pounds of sugar, and the debt of '95, 24,750,000,000 pounds of sugar, an excess of 643,348,314 pounds in favor of the debt of '66;

that the debt of '66 would purchase 181,967,213 cwt. of beef
and the debt of '95 130,263,136 cwt. of beef, an excess of
51,704,077 cwt. in favor of the debt of '66; that the debt
of '66 would purchase, approximately, 37,666,666 acres of
farming lands in the Ohio and Mississippi valleys and the
debt of '95, 35,357,142 acres, an excess in favor of the debt
of '66 of 2,309,524 acres. Thus it will be seen that the debt
of '95 would purchase of the first six articles mentioned
more than the debt of '66 would, at their respective dates,
by a large percentage, and the debt of '95 only fails to purchase as much of the last three items mentioned as the debt
of '66, by a small percentage.

---

Note 2—

An Act to Strengthen the Public Credit.

Be it enacted by the Senate and House of Representa-
tives of the United States of America, in Congress assem-
bled, that in order to remove any doubt as to the purpose
of the government to discharge all just obligations to
the public creditors, and to settle conflicting questions and
interpretations of the laws by virtue of which such obligations have been contracted, it is hereby provided and declared that the faith of the United States is solemnly
pledged to the payment in coin, or its equivalent of all
the obligations of the United States not bearing interest
known as United States notes and of all the interest-
bearing obligations of the United States, except in cases
where the law authorizing the issue of any such obligation has expressly provided that the same may be paid in
lawful money, or other currency than gold and silver. But
none of said interest-bearing obligations not already due
shall be redeemed or paid before maturity unless at such
time United States notes shall be convertible into coin at
the option of the holder, or unless at such time the bonds
of the United States bearing a lower rate of interest than
the bonds to be redeemed can be sold at par in coin.

And the United States also solemnly pledges its faith
to make provision at the earliest practicable period for
the redemption of the United States notes in coin.

Approved March 18th, 1869.

Note 3—

"This agreement, entered into this 8th day of February, 1895, between the Secretary of the Treasury of the United States, of the first part, and Messrs. August Belmont & Co., of New York, on behalf of Messrs. N. M. Rothschild & Sons, of London, England, and themselves, and J. P. Morgan & Co., of New York, on behalf of J. P. Morgan & Co., of London, and themselves, parties of the second part, witnesseth:

Whereas, It is provided by the Revised Statutes of the United States, section 3700, that the Secretary of the Treasury may purchase coin with any of the bonds or notes of the United States authorized by law at such rate and upon such terms as he may deem most advantageous to the public interests, and the Secretary of the Treasury now deems that an emergency exists in which the public interests require that as hereinafter provided coin shall be purchased with the bonds of the United States of the description hereinafter mentioned, authorized to be issued under the act entitled, 'An act to provide for the resumption of specie payments,' approved Jan. 14, 1875, being bonds of the United States described in an act of Congress approved July 14, 1870, entitled 'An act to authorize the refunding of the national debt.'

"Now, therefore, the said parties of the second part hereby agree to sell and deliver to the United States 3,500,000 ounces of standard gold coin of the United States at the rate of $17.80441 per ounce, payable in United States 4 per cent thirty-year coupon or registered bonds, said bonds to be dated Feb. 1, 1895, and payable at the pleasure of the United States after thirty years from date issued under the acts of congress of July 14, 1870, Jan. 20, 1871 and Jan. 14, 1875, bearing interest at the rate of 4 per cent per annum, payable quarterly.

First. Such purchase and sale of gold coin being made on the following conditions:

1. At least one-half of all coin delivered hereinunder shall be obtained in and shipped from Europe, but the shipment shall not be expected to exceed 500,000 ounces per month, unless the parties of the second part shall consent therto.

2. All deliveries shall be made at any of the sub-treasuries or at any other legal depository of the United States.

3. All gold coins delivered shall be secured on the basis of twenty-five and eight tenths grains of standard gold per dollar, if within limit of tolerance.

4. Bonds delivered under this contract are to be delivered free of accrued interest; which is to be assumed and paid by the parties of the second part at the time of their delivery to them.

Second. Should the Secretary of the Treasury desire to offer or sell any of the bonds of the United States on or before the first of October, 1895, he shall first offer the same to the parties of the second part; but thereafter he shall be free from every such obligation to the parties of the second part.

Third. The Secretary of the Treasury hereby reserves the right within ten days from this date, in case he shall receive authority from Congress therefor, to substitute any bonds of the United States bearing 3 per cent interest, of which the principal and interest shall be specifically payable in United States gold coin of the present weight and fineness, for the bonds herein alluded to; such 3 per cent bond to be accepted by the parties of the second part at par, i. e., at $18.60465 per ounce of standard gold.

Fourth. No bonds shall be delivered to the parties of the second part or either of them except in payment for coin from time to time received hereunder, whereupon the Secretary of the Treasury of the United States shall and will deliver the bonds as herein provided at such places as shall be designated by the parties of the second part. Any expense of delivery out of the United States shall be assumed and paid by the parties of the second part.

Fifth. In consideration of the purchase of such coin the parties of the second part and their associates hereunder assume and will bear all the expense and inevitable loss of bringing gold from Europe hereunder, and as far as lies in their power will exert all financial influence and WILL MAKE ALL LEGITMATE EFFORTS TO PROTECT THE TREASURY OF THE UNITED STATES AGAINST WITHDRAWALS OF GOLD pending the complete performance of this contract.

"In witness whereof the parties hereunto set their hands
in five parts this 8th day of February, 1895.

J. G. Carlisle,
Secretary of the Treasury.

August Belmont & Co.,
On behalf of Messrs. N. M. Rothschild & Sons, London, and themselves.

J. P. Morgan & Co.,
On behalf of J. P. Morgan & Co., London, and themselves.

Attest:W. E. Curtis, Francis Linde Stetson."

---

A Letter to One of the Free Coinage Champions from
Jay Cooke,
The Financial Agent and Advisor of the United States
During the Administration of Abraham Lincoln—The
Man Who Sold Our Bonds, at Home and Abroad, and
Lent His Great Credit and Energy to Sustain and Sup-
ply Our National Treasury with Funds to Carry on the
War of the Great Rebellion.

"Philadelphia, Sept. 3, 1895.
Chas. A. Towne, M. C.,
Duluth, Minn.:
Hon. and Dear Sir:

Some one has sent me the Duluth Herald containing
your very able and convincing speech on the silver question. I am glad to see that you are so outspoken, plain and
fearless, and I agree to every word you utter.

Please send me some copies of the Herald containing
the speech in full, and if you have it in pamphlet form I
will subscribe and circulate some of the pamphlets.

I enclose $2.00 for copies of Herald.

Yours sincerely,
Jay Cooke."